Geometry
GRADE 7

Written by
Theresa Kane McKell

Illustrated by
Patricia Grush and
Robin DeWitt

Cover Illustration
by Susan Cumbow

FS112033 Geometry Grade 7
All rights reserved—Printed in the U.S.A.

Copyright © 1998 Frank Schaffer Publications, Inc.
23740 Hawthorne Blvd.
Torrance, CA 90505

TABLE OF CONTENTS

INTRODUCTION

This seventh grade geometry book is part of the *Math Minders* series. It will provide your math students with an appealing variety of fun and exciting math activities. As you teach math throughout the seventh grade year, activities can be found in this book covering important geometric topics—from identifying and drawing different figures and their parts to calculating the perimeter, area, surface area, and volume of two-dimensional and three-dimensional shapes.

Most of the pages in this book provide an exciting puzzle to practice a specific skill. Along with the skills they learn, students will enjoy other interesting facts, riddles, and messages relating to the real world. These activities will be extremely useful for your seventh grade math students.

The activities are laid out by specific concepts. The table of contents provides a broad overview of the page titles and topics addressed. To find a specific skill, read the top right-hand corner of each page, where the skills are listed.

Most geometric skills throughout this book are introduced with one or two activities for practice. The first page on a topic includes the formulas needed to work each problem. Consecutive pages on a given topic will be increasingly more difficult. Carefully preview each activity so you are certain students are doing the appropriate activity for their skill level. The solutions to these activities are in the answer key at the back of the book.

Very often students are not very excited about math. Most activities in this book are not only about geometry. As students complete the activities, they become aware that there is more than one lesson, fact, or riddle addressed on each page. Take this opportunity to make geometry topics fun, relevant, and interesting to your students.

Geometry

GRADE **7**

Name____

CAN YOU NAME IT?

Lines, line segments, rays, angles, planes

Using symbols, name the following drawings in as many ways as possible.

A.

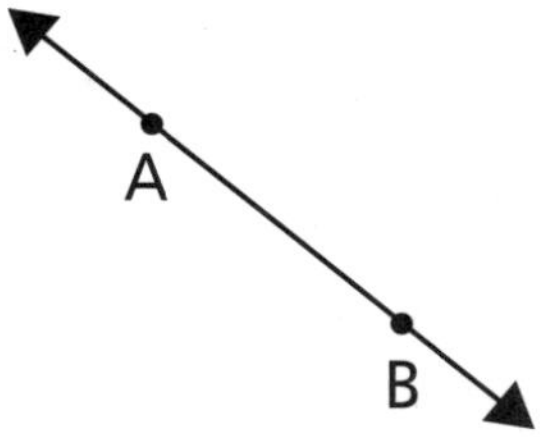

B.

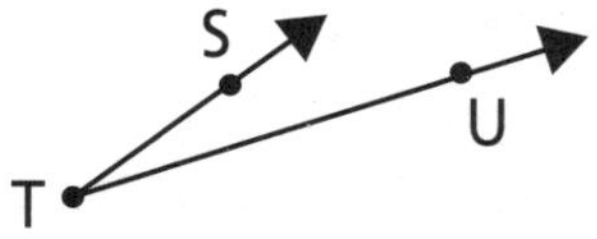

C.

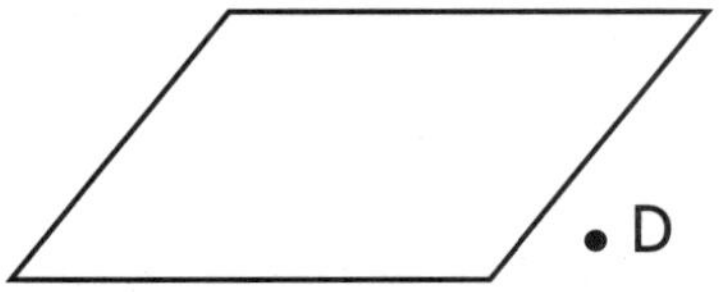

D.

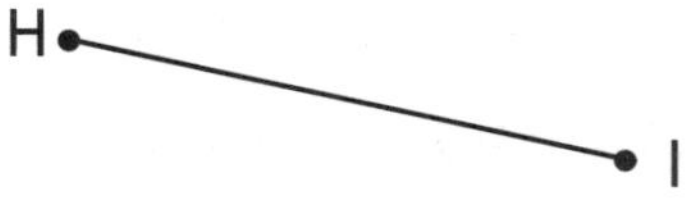

E.

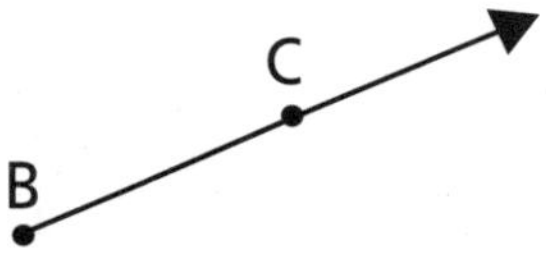

F.

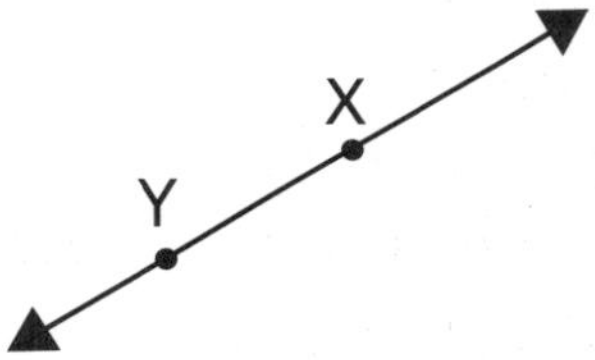

G.

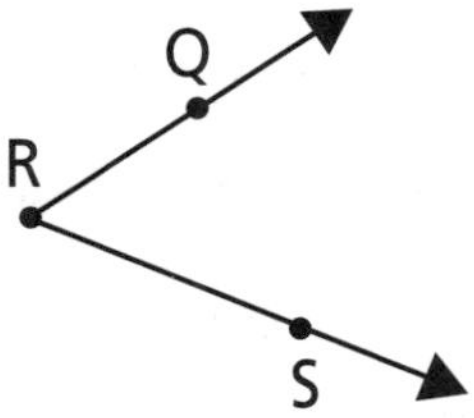

H.

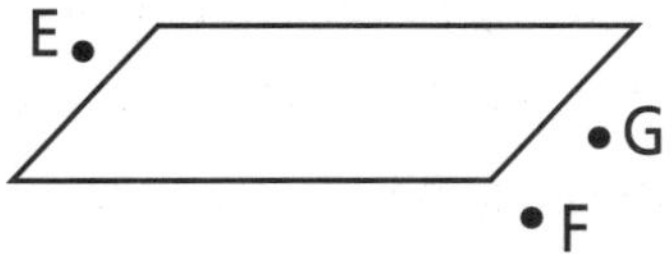

I.

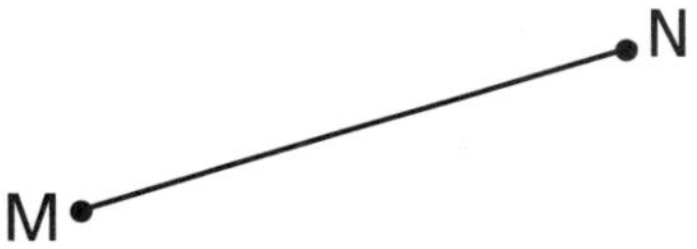

J.

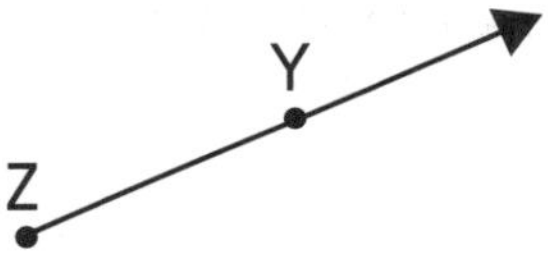

DOODLING WITH DRAWINGS··············

Make and label a drawing for each of the problems below.

A. Line ST intersects Line RV at point B.

B. Ray AB is parallel to line segment XY.

C. Line EF is perpendicular to both parallel lines CD and QR at points A and S, respectively.

D. Line segment UV is in the plane MNO.

E. Lines GH and VW are skew lines.

F. Line segments KL and OP are on line XY.

G. Ray ZW intersects line UN at point C in the plane T.

H. Line segment XZ is perpendicular to line segment AB in the plane EFG.

I. Ray AD contains the segments AF, AW, ED, and CD.

J. Line WX and line UV are skew lines.

Name_____

Use the drawing below to answer the following questions.

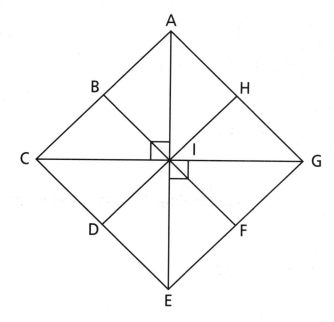

A. Name four pair of adjacent angles.

_____ _____

_____ _____

B. Name four pair of vertical angles.

_____ _____

_____ _____

C. If ∠AIH = 45°, what does ∠DIE equal?

_____

D. Name the adjacent angles to ∠EIF.

_____ _____

E. The sum of ∠AIH and ∠AIB is equal to the sum of what two angles?

_____

_____

F. If ∠HIF = 90°, what is the measure of ∠DIB?

_____

G. Name the vertical angle and adjacent angles of ∠BIC.   Vertical angle = _____

Adjacent angles = _____

H. The sum of ∠BIC and ∠HIG is equal to the sum of what two angles?

_____

_____

MEASURING UP TO CLASS.............

Measure each angle below with your protractor. Classify each angle as acute, obtuse, or right. Write the angle measure on the blank inside the angle and its classification on the blank next its letter.

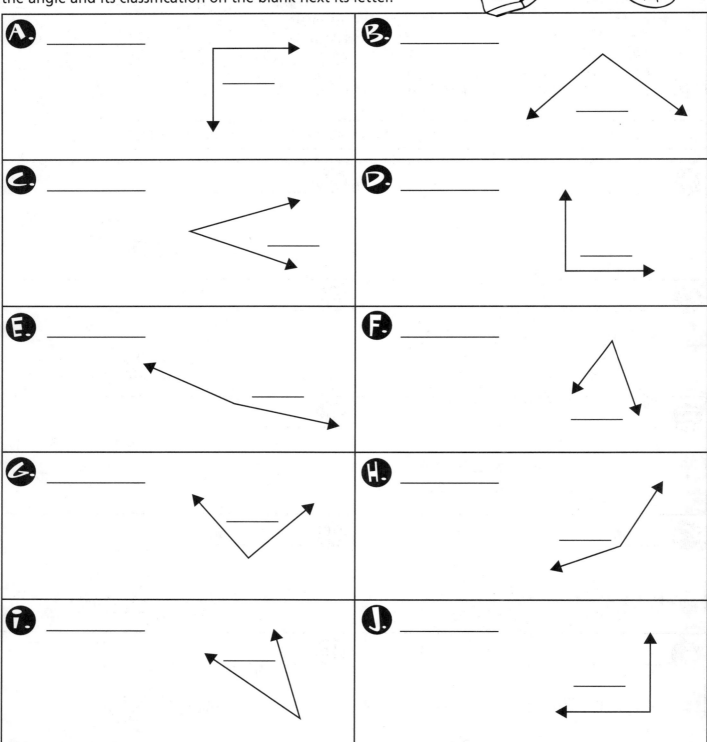

JOIN THE CROWD

Find the measure of the missing angles and write them on the lines by the angles. Find the letter that represents each angle measure and write it in the space next to its letter. Read the letters to find the answer to the following fun fact.

(L)	80°	**(T)**	40°	**(A)**	90°	**(O)**	15°
(G)	30°	**(F)**	45°	**(S)**	150°	**(Y)**	55°
(C)	60°	**(R)**	70°	**(I)**	25°	**(A)**	135°
(I)	75°	**(E)**	65°	**(N)**	50°	**(M)**	85°

Which state has more people than any other state in the United States?

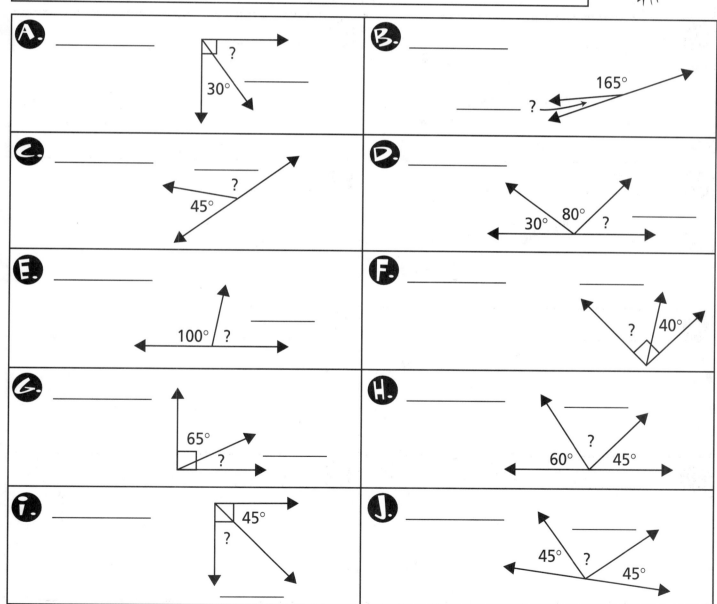

THE BULL-YGONS

In 1994, Chicago unveiled a bronze statue of this basketball star and his #23 jersey was retired. Who is the famous person?

To find the answer to this question, classify each of the following shapes using your knowledge of convex, concave, and regular polygons. Find the letter on the right of each problem that represents the correct name of the shape. Place the letter in the blank at the bottom of the page above its problem letter.

A. (S) Concave polygon   (M) Convex polygon

B. (I) Regular polygon   (C) Not a regular polygon

C. (C) Not a polygon   (T) Regular polygon

D. (T) Convex polygon   (H) Concave polygon

E. (A) Regular, convex polygon   (I) Not regular, concave polygon

F. (Y) Concave polygon   (E) Not a polygon

G. (E) Regular, concave polygon   (L) Convex polygon

H. (P) Convex polygon   (J) Not a polygon

I. (I) Convex, regular polygon   (O) Convex polygon

J. (P) Concave polygon   (R) Not a polygon

K. (D) Regular, convex polygon   (P) Not regular, concave polygon

L. (E) Concave polygon   (A) Not a polygon

M. (S) Not a polygon   (N) Concave polygon

___ ___ ___ ___ ___ ___ ___   ___ ___ ___ ___ ___ ___
 A   B   C   D   E   F   G   H   I   J   K   L   M

POLYGON PARADE

Below is a parade of polygons. Fill in the chart with the name (according to its number of sides), number of sides, number of angles, and number of diagonals of each polygon participating in this parade.

Polygons	Name	Number of		
		Sides	Angles	Diagonals
A. Float #1				
B. Float #2				
C. Float #3				
D. Float #4				
E. Float #5				
F. Float #6				

TUMMY-N-TRIANGLE TIME........... Classifying triangles using their sides and their angles

What is one of the favorite foods of many kids in America?

To find the answer to this question, classify each of the following triangles according to its given measurements of either its sides or its angles. Find the letter on the right of each problem that represents the answer. Write the letter above its problem letter in the correct blank at the bottom of the page.

A. 3 cm, 6 cm, 9 cm
(C) scalene
(D) isosceles
(H) equilateral

B. 45°, 90°, 45°
(A) obtuse
(H) right
(U) acute

C. 60°, 35°, 85°
(T) scalene
(Y) obtuse
(E) acute

D. 10 in, 8 in, 10 in
(E) isosceles
(F) equilateral
(W) scalene

E. 9m, 9m, 9m
(Q) obtuse
(S) equilateral
(P) scalene

F. 120°, 40°, 20°
(V) scalene
(T) acute
(E) obtuse

G. 14 cm, 7 cm, 15 cm
(B) scalene
(D) right
(O) isosceles

H. 55°, 90°, 35°
(G) acute
(U) right
(H) obtuse

i. 40°, 100°, 40°
(R) obtuse
(T) isosceles
(J) acute

J. 15cm, 13 cm, 13 cm
(I) equilateral
(L) acute
(G) isosceles

K. 75°, 35°, 70°
(E) acute
(B) isosceles
(C) right

L. 3 m, 3 m, 3 m
(S) obtuse
(R) equilateral
(N) scalene

M. 41°, 130°, 9°
(M) acute
(X) equilateral
(A) obtuse

N. 25 in, 31 in, 35 in
(N) scalene
(G) acute
(S) equilateral

O. 50°, 50°, 80°
(J) equilateral
(D) acute
(O) obtuse

P. 2 ft, 2 ft, 2 ft
(F) equilateral
(Y) isosceles
(E) obtuse

Q. 17°, 62°, 101°
(W) acute
(G) right
(R) obtuse

R. 45 mm, 19 mm, 45 mm
(I) isosceles
(T) acute
(U) equilateral

S. 28°, 62°, 90°
(V) obtuse
(E) right
(H) scalene

T. 16 cm, 24 cm, 42 cm
(S) scalene
(P) isosceles
(W) acute

__ __ __ __ __ __ __ __ __ __ __ __
A B C D E F G H I J K L

__ __ __ __ __ __ __ __ !!!
M N O   P Q R S T

10

Name_____

Make and label a drawing of each of the following.

A. △ABC is obtuse and scalene.

B. △IJK is an acute, equilateral triangle.

C. △XYZ is an isosceles, right triangle.

D. △RST is obtuse and isosceles.

E. △PQR is an acute, scalene triangle.

F. △EFG is right and scalene.

G. △MNO is an isosceles, acute triangle.

Name_____

What are the properties of each of the quadrilaterals below? Find the letters on the right that represent the shape's properties. Write the letters in the blanks next to each problem. Letters may be used more than once.

A. Parallelogram _____ _____ _____

B. Trapezoid _____ _____

C. Square _____ _____ _____

_____ _____ _____

D. Rhombus _____ _____ _____ _____ _____

E. Rectangle _____ _____ _____ _____

(A)	**This shape is a quadrilateral.**
(B)	**All four sides are congruent.**
(C)	**This shape has only one pair of parallel sides.**
(D)	**Opposite sides are parallel.**
(E)	**Opposite sides are congruent.**
(F)	**Opposite angles are congruent.**
(G)	**All four angles are congruent.**

Using the properties of the quadrilaterals you found above, tell whether each of the following statements is true or false. If the statement is false, state your reason why.

F. A parallelogram is a quadrilateral. _____ _____

G. All squares and rectangles are parallelograms._____ _____

H. A rhombus is a square._____ _____

I. A square is a rhombus._____ _____

J. A trapezoid is parallelogram._____ _____

K. A square is a rectangle._____ _____

L. A rectangle is a square._____ _____

M. A trapezoid is a quadrilateral._____ _____

N. A rhombus is a rectangle._____ _____

O. All quadrilaterals are parallelograms._____ _____

_____

TOOLIN' WITH TRIANGLES..................

You can draw diagonals from one vertex in any regular polygon to divide the shape into triangles. You also know that the sum of the angles in any triangle is 180°. Given each of the regular polygons below, fill in the chart with the number of sides, number of angles, number of triangles, total angle measure, and the measure of each angle.

Polygons	Number of			Total Angle Measure	Measure of Each Angle
	Sides	Angles	Triangles		
A. Triangle					
B. Quadrilateral					
C. Pentagon					
D. Hexagon					
E. Octagon					
F. Decagon					
G. N-gon					

Name_____

OUT iN SPACE ·············

From where in Florida did the first space satellite, Explorer I, and the first manned spacecraft, Apollo II, launch?

To find the answer to this historical fact, find the perimeter of each of the polygons. Find the letter that matches each perimeter at the bottom of the page. Write the letter next to its problem. Read the letters to find the answer.

A. _____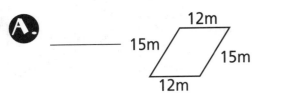

B. _____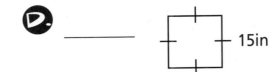

C. _____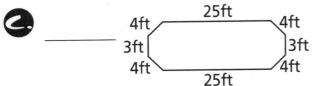

D. _____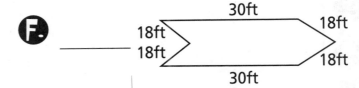

E. _____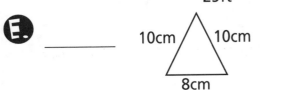

F. _____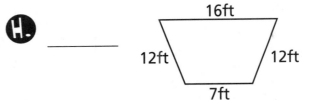

G. _____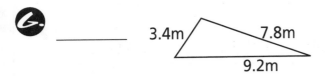

H. _____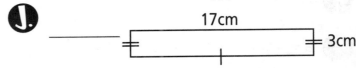

i. _____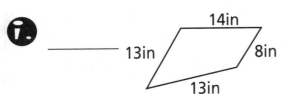

J. _____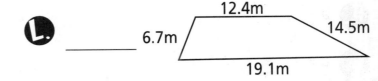

K. _____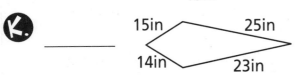

L. _____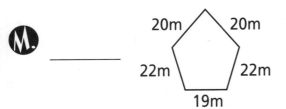

M. _____

(V)	60 in	(A)	72 ft	(N)	103 m
(P)	28 cm	(E)	132 ft	(E)	20.4 m
(A)	88 ft	(H)	42 m	(L)	52.7 m
(R)	47 ft	(K)	16 ft	(M)	75 in
(A)	40 cm	(C)	48 in	(N)	126 cm
(C)	54 m	(O)	84 cm	(A)	77 in

FS112033 Geometry Grade 7

THiS LiTTLE PiGGY.....................

What position did Perry the Pig play on the baseball team?

To find the answer to this riddle, find the perimeter of each of the following shapes. In questions A-C, find the perimeter of each square. In questions D-F, find the perimeter of each rectangle. In questions G-I, find the perimeter of each triangle. Find the letter on the right that represents the correct answer. Write each letter above its problem number at the bottom of the page.

A. s = 75 m   (P) 150 m   (T) 200 m   (S) 300 m

B. s = 18 in   (H) 72 in   (I) 324 m   (E) 90 m

C. s = 38.2 cm   (J) 76.4 cm   (O) 152.8 cm   (G) 1459.24 cm

D. l = 45 m, w = 30 m   (R) 150 m   (G) 1350 m   (L) 75 m

E. l = 15.3 cm, w = 12.7 cm   (Y) 194.31 cm   (O) 28 cm   (T) 56 cm

F. l = 78 in, w = 52 in   (U) 130 in   (S) 260 in   (H) 4056 in

G. a = 4.5 m, b = 9.2 m, c = 12.1 m   (L) 25.8 m   (W) 13.7 m   (P) 27.7 m

H. a = 13 in, b = 21 in, c = 34 in   (B) 25 in   (E) 70 in   (O) 68 in

I. a = 44 cm, b = 58 cm, c = 71 cm   (F) 115 cm   (P) 173 cm   (D) 129 cm

___ ___ ___ ___ ___ _ ___ ___ ___ ___!
 A   B   C   D   E   F   G   H   I

Name_____

To find the circumference of a circle, you can either use C = π•d or C = 2•π•r. Find the circumference of each of the cookies below. Place each answer in the blank next to its cookie. Use 3.14 for the value of π.

A.  _____ d = 7 cm

B.  _____ r = 16 mm

C.  _____ d = 25 mm

D.  _____ r = 2 cm

E.  _____ d = 11 cm

F.   _____ r = 3 cm

G.  _____ d = 45 mm

H.  _____ r = 4 cm

I.  _____ d = 9 cm

J. 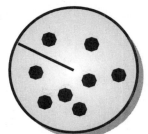 _____ r = 7 cm

No SMARTY PANTS

What happened to Mr. Pants when he cut school?

To find the answer to this riddle, calculate the circumference in each problem below using either C = π•d or C = 2•π•r. Find the letter on the right that represents each answer. Write the letter above its problem letter at the bottom of the page. Use 3.14 for π. Round each answer to the nearest tenth.

A. d = 12 in   (R) 37.8 in   (H) 37.7 in   (G) 38.0 in

B. r = 18 ft   (E) 113.0 ft   (I) 113.1 ft   (K) 113.21 ft

C. d = 2.5 cm   (N) 8.0 cm   (W) 7.9 cm   (L) 7.8 cm

D. r = 1.4 in   (F) 8.9 in   (A) 8.8 in   (J) 9.0 in

E. d = 15 ft   (O) 47.0 ft   (U) 46.9 ft   (S) 47.1 ft

F. r = 9 m   (S) 56.5 m   (B) 56.2 m   (C) 56.0 m

G. d = 6.7 in   (A) 21.1 in   (U) 21.0 in   (M) 21.2 in

H. r = 10.2 ft   (N) 64.21 ft   (R) 64.01 ft   (S) 64.1 ft

I. d = 21 cm   (P) 65.9 cm   (H) 65.8 cm   (F) 66.0 cm

J. r = 25 mm   (E) 157.0 mm   (I) 157.21 mm   (D) 157.1 mm

K. d = 17.1 in   (L) 54.0 in   (N) 53.7 in   (V) 53.6 in

L. r = 3.8 ft   (C) 24.0 ft   (S) 23.8 ft   (D) 23.9 ft

M. d = 5.6 m   (E) 17.6 m   (B) 17.7 m   (U) 18.0 m

N. r = 20 cm   (G) 126.0 cm   (R) 125.6 cm   (T) 125.5 cm

O. d = 6.9 in   (I) 21.8 in   (D) 21.0 in   (E) 21.7 in

P. r = 4.7 ft   (D) 29.5 ft   (M) 30.0 ft   (W) 29.6 ft

___ ___   ___ ___ ___
A   B   C   D   E

___ ___ ___ ___ ___ ___ ___ ___ ___ ___ ___!!
F   G   H   I   J   K   L   M   N   O   P

Name_____

What does an egg do when another egg bothers it?

To find the answer to this egg-beater question, sketch each figure and label its base and height. Calculate the area of the parallelograms (A = b h) and triangles (A = ½ b h). Find the letter at the bottom of the page that represents each answer. Write the letter next to its problem number. Read down to find the answer.

A. _____ A parallelogram has a base of 12 in and a height of 9 in.

B. _____ A triangle has a base of 9 m and a height of 10 m.

C. _____ The length of a side of a square is 8 ft long.

D. _____ A triangle has a base of 8 cm and a height of 14 cm.

E. _____ A rectangle has a base of 24 mm and a height of 15 mm.

F. _____ A triangle has a base of 18 in and a height of 25 in.

G. _____ A parallelogram has a base of 20 m and a height of 35 m.

H. _____ A triangle has a base of 30 ft and a height of 50 ft.

I. _____ A square has a height of 11 cm.

J. _____ A triangle has a base of 22 mm and a height of 40 mm.

(G) 64 ft²	**(S) 750 ft²**	**(T) 440 mm²**	**(U) 22 cm²**
(H) 72 in²	**(L) 16 ft²**	**(N) 56 cm²**	**(E) 108 in²**
(R) 225 in²	**(E) 700 m²**	**(M) 80 ft²**	**(O) 360 mm²**
(P) 18 m²	**(G) 45 m²**	**(I) 121 cm²**	**(W) 55 m²**

Name_____

NAME THAT STATE · · · · · · · · · · · · · · · · · ·

What is the only state in the U.S. that was named for a president?

To find the answer to this interesting question, calculate the area of the trapezoids using the formula $A = \frac{1}{2}(b1 + b2)h$. Write each answer in the blank next to each shape's letter. Match the letter of each problem with its area at the bottom of the page.

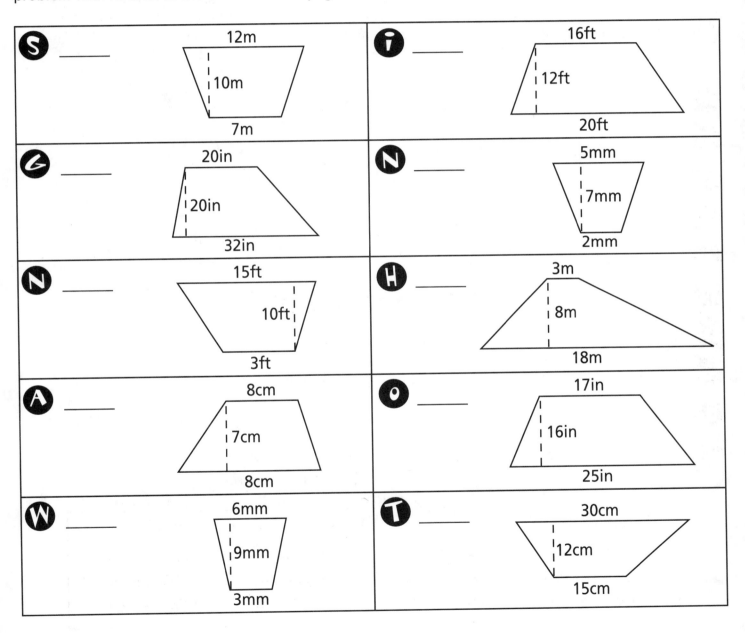

S ____ 12m / 10m / 7m

i ____ 16ft / 12ft / 20ft

G ____ 20in / 20in / 32in

N ____ 5mm / 7mm / 2mm

N ____ 15ft / 10ft / 3ft

H ____ 3m / 8m / 18m

A ____ 8cm / 7cm / 8cm

O ____ 17in / 16in / 25in

W ____ 6mm / 9mm / 3mm

T ____ 30cm / 12cm / 15cm

____ 40.5 mm² ____ 56 cm² ____ 95 m² ____ 84 m² ____ 216 ft² 24.5 mm² 520 in² 270 cm² 336 in² 90 ft²

PIZZA PLEASER ·

Find the area of each of the pizzas using the formula $A = \pi r^2$. Write each answer in the blank next to the problem. Use 3.14 for π.

A. _____ A small sausage, mushroom, and cheese pizza with a radius of 5 inches

B. _____ A large pizza with pepperoni, hamburger, and extra cheese with a radius of 8 inches

C. _____ A medium taco pizza with a radius of 7 inches

D. _____ An extra large deluxe pizza with a radius of 10 inches

E. _____ A personal pizza with only cheese with a diameter of 6 inches

F. _____ An extra large pizza with the "works" with a diameter of 30 inches

G. _____ A family-style pizza with everything on it with a diameter of 18 inches

H. _____ A student-style pizza with extra meat and cheese with a diameter of 24 inches

I. _____ You and your friends went out for pizza. You ordered the "Monster" pizza with a radius of 16 inches. You and your friends could only eat half of the pizza. What was the area of the pizza that was left?

J. _____ Your class is having a pizza party. You are going to order 12 pizzas, each having a diameter of 15 inches. Assuming you will eat your entire order, what is the total area of the pizzas your class will eat?

Name_____

Why can't the geometry teacher walk to school?

To find the answer to this amusing riddle, find the area of each of the figures. Write each answer in the space next to its shape's letter. Match each letter with its area below.

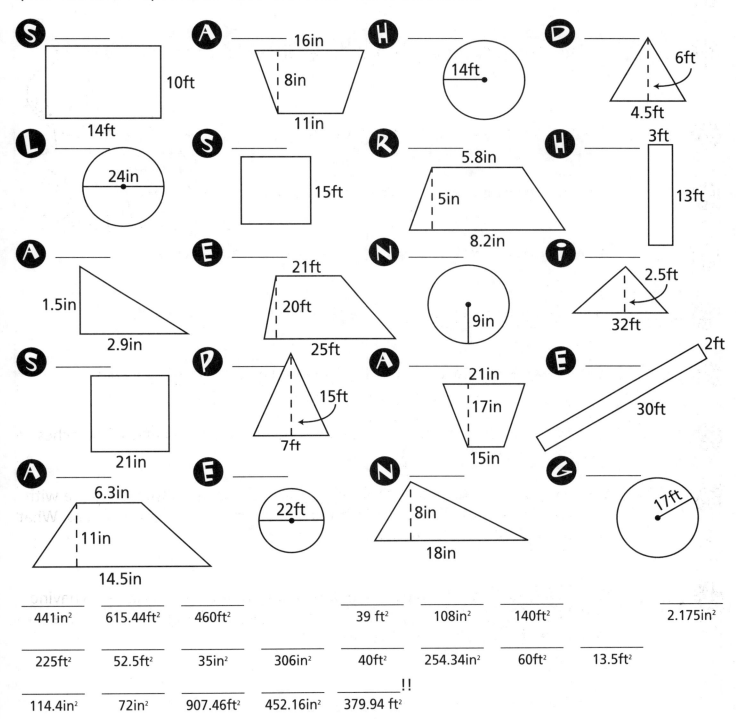

| 441in² | 615.44ft² | 460ft² | | 39 ft² | 108in² | 140ft² | | 2.175in² |

| 225ft² | 52.5ft² | 35in² | 306in² | 40ft² | 254.34in² | 60ft² | 13.5ft² |

| 114.4in² | 72in² | 907.46ft² | 452.16in² | 379.94 ft² !! |

FS112033 Geometry Grade 7

SYMMETRY SEARCHIN'..................

Tell whether each of the following shapes has a line of symmetry and/or a turn of symmetry. Write yes or no in the blanks provided. If yes, draw all lines of symmetry and/or identify the angle measure of each turn.

A.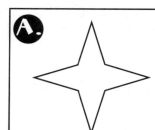
line of symmetry _____
turn of symmetry _____
if so, ∠m = _____

B.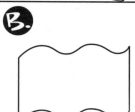
line of symmetry _____
turn of symmetry _____
if so, ∠m = _____

C.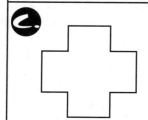
line of symmetry _____
turn of symmetry _____
if so, ∠m = _____

D.
line of symmetry _____
turn of symmetry _____
if so, ∠m = _____

E.
line of symmetry _____
turn of symmetry _____
if so, ∠m = _____

F.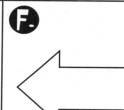
line of symmetry _____
turn of symmetry _____
if so, ∠m = _____

G.
line of symmetry _____
turn of symmetry _____
if so, ∠m = _____

H.
line of symmetry _____
turn of symmetry _____
if so, ∠m = _____

I.
line of symmetry _____
turn of symmetry _____
if so, ∠m = _____

J.
line of symmetry _____
turn of symmetry _____
if so, ∠m = _____

Name_____

Complete the table below.

	Polyhedron	Sketch of Shape	Number of		
			Faces	Vertices	Edges
A.	Triangular Prism				
B.	Triangular Pyramid				
C.	Rectangular Prism				
D.	Rectangular Pyramid				
E.	Pentagonal Prism				
F.	Pentagonal Pyramid				
G.	Hexagonal Prism				
H.	Hexagonal Pyramid				
I.	Octagonal Prism				
J.	Octagonal Pyramid				

FS112033 Geometry Grade 7

Name_____

What does the funniest kid in class eat for breakfast?

To find the answer to this riddle, circle the letter of the answer on the right of each polyhedron. Write the answer letter above its problem letter at the bottom of the page.

  (W) Rectangular pyramid   (C) Rectangular prism

  (R) Triangular prism   (I) Rectangular prism

  (U) Cone   (E) Cylinder

  (A) Sphere   (N) Circular prism

  (O) Triangular pyramid   (M) Cone

  (B) Pentagonal prism   (O) Hexagonal Prism

  (F) Rectangular pyramid   (R) Triangular pyramid

  (W) Cube   (T) Rectangular pyramid

  (K) Triangular cone   (I) Triangular pyramid

 (T) Octagonal prism   (S) Decagonal prism

___ ___ ___ ___ ___   ___ ___   ___ ___ ___!!
A   B   C   D   E   F   G   H   I   J

iCE CREAM iNCiDENT ....

What would you get if you dropped your ice cream on the ground?

To find the answer to this riddle, find the surface area of each of the following prisms. Use the formula SA = area of faces + area of bases. Write each answer in the blank next to its shape. Match each letter with its area at the bottom of the page.

i _____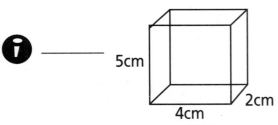

E _____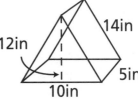

P _____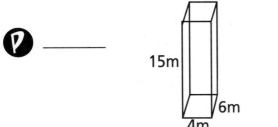

S _____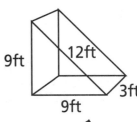

A _____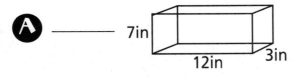

L _____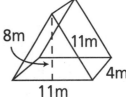

P _____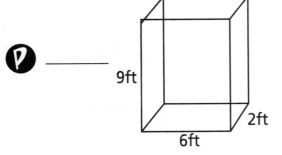

C _____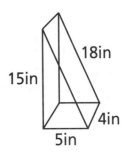

L _____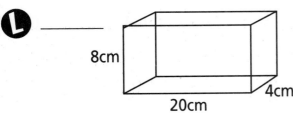

O _____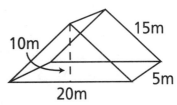

___ ___ ___ ___ ___ ___ ___ ___ ___ ___!

| 282 in² | | 348 m² | 544 cm² | 450 m² | 168 ft² | 171 ft² | 76 cm² | 242 in² | 220 m² | 310 in² |

Name_____

Find the surface area of the pyramids described. Use the formula
SA = area of faces + area of base. Sketch each pyramid and label
its dimensions. Write each answer in the blank.

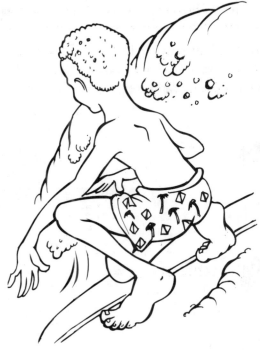

A. _____ A pyramid with a base of 4 cm by 4 cm
and an altitude of 6 cm

B. _____ A pyramid with a base of 8 cm by 8 cm
and an altitude of 16 cm

C. _____ A pyramid with a base of 10 cm by 10 cm
and an altitude of 8 cm

D. _____ A pyramid with a base of 9 cm by 9 cm
and an altitude of 18 cm

E. _____ A pyramid with a base of 12 cm by 12 cm
and an altitude of 20 cm

F. _____ A pyramid with a base of 15 cm by 15 cm
and an altitude of 25 cm

G. _____ A pyramid with a base of 11 cm by 11 cm
and an altitude of 21 cm

H. _____ A pyramid with a base of 19.1 cm by 19.1 cm
and an altitude of 30 cm

I. _____ A pyramid with a base of 30.7 cm by 30.7 cm
and an altitude of 35.5 cm

J. _____ A pyramid with a base of 22.2 cm by 22.2 cm
and an altitude of 28.8 cm

Name_____

When do candles like to party?

To find the answer to this riddle, find the surface area of each of the cylinders below. Use the formula SA = 2πrh + 2πr². Circle the letter on the right of each problem that represents its answer. Write the letters on the answer line at the bottom of the page. Use 3.14 for π.

A. 2in / 5in
(O) 87.92 in²
(I) 90.0 in²
(P) 3244.8 in²

B.  5in / 8in
(R) 418.3 in²
(S) 232.41 in²
(N) 408.2 in²

C. 4in / 3in
(H) 185.43 in²
(W) 175.84 in²
(O) 196.71 in²

D.  10in / 15in
(I) 1570 in²
(U) 1562 in²
(L) 1579 in²

E. 15in / 35in
(T) 4762 in²
(E) 4750.3 in²
(C) 4710 in²

F.  30in / 20in
(K) 9420 in²
(A) 9210 in²
(J) 908.45 in²

G. 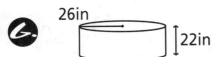 26in / 22in
(D) 7564.32 in²
(E) 7837.44 in²
(Y) 9872.1 in²

H. 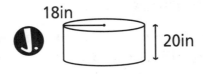 8in / 14in
(N) 1105.28 in²
(I) 1109.43 in²
(K) 1124.8 in²

I. 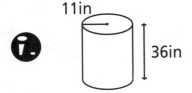 11in / 36in
(M) 3250 in²
(D) 3246.76 in²
(H) 9235.1 in²

J. 18in / 20in
(B) 4290.1 in²
(P) 4300 in²
(S) 4295.52in²

___ ___   ___ ___ ___ ___ ___ ___ ___ ___!!
 A   B   C   D   E   F   G   H   I   J

Name____

SURFACE AREA STYLE

Surface area of prisms, pyramids, and cylinders

Find the surface area of each of the following figures. Sketch each shape and label its dimensions. Use 3.14 for π. Write each answer in the blank next to its problem.

Sketch your shapes here.

A. ________ A cylinder with a radius of 9 ft and a height of 15 ft

B. ________ A rectangular prism with a length of 10 cm, a height of 12 cm, and a width of 5 cm

C. ________ A pyramid with a square base of 4 m by 4 m and an altitude of 11 m

D. ________ A cylinder with a radius of 15 in and a height of 20 in

E. ________ A pyramid with a square base of 14 ft by 14 ft and an altitude of 16 ft

F. ________ A rectangular prism with a length of 40 cm, a height of 30 cm, and a width of 20 cm

G. ________ A cylinder with a radius of 25 m and a height of 13 m

H. ________ A rectangular prism with a length of 5 in, a height of 7 in, and a width of 9 in

I. ________ A pyramid with a square base of 25 cm by 25 cm and an altitude of 30 cm

J. ________ A rectangular prism with equilateral bases with sides 10 cm in length and an altitude of 14 cm

ROLLIN' IN THE DOUGH.................

Volume of prisms and cylinders

What is the name of this American family who is found to be the richest family in the world?

To find the answer to this interesting fact, find the volume of the prisms (V = lwh) and cylinders (V = ($\pi r^2 h$). Write each answer in the blank next to its shape. Match each letter with its volume at the bottom of the page. Use 3.14 for π.

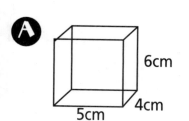

A 6cm 4cm 5cm

_____

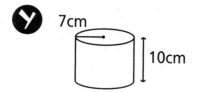

Y 7cm 10cm

_____

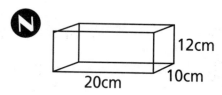

N 12cm 20cm 10cm

_____

F 3cm 6cm

_____

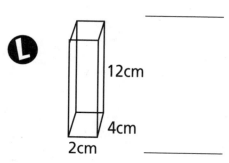

L 12cm 4cm 2cm

_____

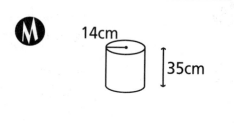

M 14cm 35cm

_____

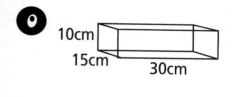

O 10cm 15cm 30cm

_____

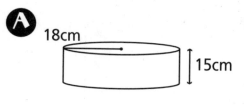

A 18cm 15cm

_____

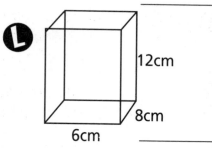

L 12cm 8cm 6cm

_____

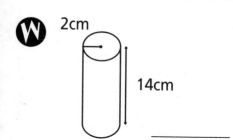

W 2cm 14cm

_____

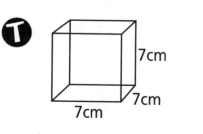

T 7cm 7cm 7cm

_____

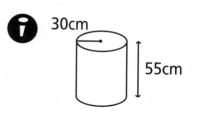

I 30cm 55cm

_____

175.84 cm³	120 cm³	96 cm³	343 cm³	4500 cm³	2400 cm³
169.56 cm³	15260.4 cm³	21540.4 cm³	155430 cm³	576 cm³	1538.6 cm³

Name_____

What kind of feet do all mathematicians have?

To find the answer to this riddle, find the volume of the rectangular pyramids (V = ⅓wh), triangular pyramids (V = ⅓ area of triangular base h), and cones (V = ⅓(πr²h)). Circle the letter on the right of each problem that represents its volume. Write each letter above its problem letter at the bottom of the page. Use 3.14 for π.

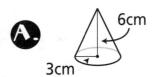

A. 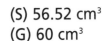 6cm / 3cm

(S) 56.52 cm³
(G) 60 cm³
(I) 92.31 cm³

B. 10cm / 6cm / 8cm

(A) 96.1 cm³
(Q) 80 cm³
(O) 42.31 cm³

C.  20cm / 12cm

(N) 3,050 cm³
(U) 3,014.4 cm³
(W) 301.4 cm³

D.  12cm / 5cm / 5cm

(S) 99.21 cm³
(M) 101.2 cm³
(A) 100 cm³

E.  15cm / 8cm

(R) 1,004.8 cm³
(T) 1,400.5 cm³
(U) 140.8 cm³

F. 15cm / 11cm / 11cm

(B) 1511.6 cm³
(E) 302.5 cm³
(C) 15,100 cm³

G. 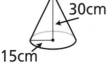 30cm / 15cm

(C) 765 cm³
(G) 7650 cm³
(F) 7,065 cm³

H.  22cm / 12cm / 12cm

(F) 5,675.1cm³
(E) 1,056 cm³
(H) 1,060 cm³

I.  36cm / 20cm

(E) 15,072 cm³
(I) 9,405.3 cm³
(K) 1575 cm³

J.  33cm / 10cm / 10cm

(D) 110 cm³
(J) 1,011 cm³
(T) 1,100 cm³

___ ___ ___ ___ ___ ___   ___ ___ ___ ___!!
 A   B   C   D   E   F   G   H   I   J

ANSWERS

Page 3

A. $\overleftrightarrow{AB}$, $\overleftrightarrow{BA}$

B. ∠STU, ∠UTS, ∠T

C. Plane D

D. $\overline{HI}$, $\overline{IH}$

E. $\overrightarrow{BC}$

F. $\overleftrightarrow{XY}$, $\overleftrightarrow{YX}$

G. ∠QRS, ∠SRQ, ∠R

H. plane F, plane G, plane EGF, plane E, plane FGE, plane EFG, plane GEF, plane FEG, plane GFE

I. $\overline{NM}$, $\overline{MN}$

J. $\overrightarrow{ZY}$

Page 4

See students' work.

A.

B.

C.

D.

E.

F. XKLOPY

G.

H.

I. EFWCD

J.

Page 5

A. Answers will vary.

B. Answers will vary.

C. 45°

D. ∠EID, ∠FIG

E. ∠EID, ∠EIF

F. 90°

G. vertical angle = ∠GIF, adjacent angles = ∠AIB, ∠CID

H. ∠GIF, ∠CID

Page 6

A. 90°, right

B. 104°, obtuse

C. 34°, acute

D. 90°, right

E. 175°, obtuse

F. 54°, acute

G. 90°, right

H. 150°, obtuse

I. 34°, acute

J. 90, right

Page 7

CALIFORNIA

Page 8

MICHAEL JORDAN

Page 9

A.	triangle	3	3	0
B.	quadrilateral	4	4	2
C.	pentagon	5	5	5
D.	hexagon	6	6	9
E.	octagon	8	8	20
F.	decagon	10	10	35

Page 10

CHEESEBURGER AND FRIES!

Page 11

See students' work.

A. 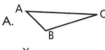 B.

C.   D.

E.   F.

G.

Page 12

Part 1

A. A, D, F

B. A, C

C. A, B, D, E, F, G

D. D, A, E, B, F

E. A, D, E, F, G

Part 2

F. T

G. T

H. F

I. T

J. F

K. T

L. F

M. T

N. F

O. F

Page 13

A. 3; 3; 1; 180°; 60°

B. 4; 4; 2; 360°; 90°

C. 5; 5; 3; 540°; 108°

D. 6; 6; 4; 720°; 120°

E. 8; 8; 6; 1,080°; 135°

F. 10; 10; 8; 1,440°; 144°

G. n; n; n − 2; (n − 2) × 180°; (n − 2) × 180° ÷ n

Page 14

CAPE CANAVERAL

Page 15

SHORT-SLOP

Page 16

A. 21.98 cm

B. 100.48 mm

C. 78.5 mm

D. 12.56 cm

E. 34.54 cm

F. 18.84 cm

G. 141.3 mm

H. 25.12 cm

I. 28.26 cm

J. 43.96 cm

Page 17

HE WAS SUSPENDERED!

Page 18

EGGNORES IT

Page 19

WASHINGTON

Page 20

A. 78.5 in²

B. 200.96 in²

C. 153.86 in²

D. 314 in²

E. 28.26 in²

F. 706.5 in²

G. 254.34 in²

H. 452.16 in²

I. 401.92 in²

J. 2,119.5 in²

Page 21

SHE HAS A SPRAINED ANGLE!!

Page 22

A. yes
   yes
   90°

B. no
   yes
   180°

C. yes
   yes
   90°

D. yes
   no

E. yes
   no

F. yes
   yes
   180°

G. yes
   no

H. yes
   yes
   45°

I. yes
   no

J. yes
   yes
   180°

Page 23

	Polyhedron	Sketch of Shape	Number of Faces	Number of Vertices	Number of Edges
A	Triangular Prism	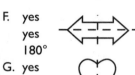	5	6	9
B	Triangular Pyramid		4	4	6
C	Rectangular Prism		6	8	12
D	Rectangular Pyramid		5	5	8
E	Pentagonal Prism		7	10	15
F	Pentagonal Pyramid		6	6	10
G	Hexagonal Prism		8	12	18
H	Hexagonal Pyramid		7	7	12
I	Octagonal Prism		10	16	24
J	Octagonal Pyramid		9	9	16

Page 24

CREAM OF WIT

Page 25

A PLOPSICLE

Page 26

A. 64 cm²
B. 320 cm²
C. 260 cm²
D. 405 cm²
E. 624 cm²
F. 975 cm²
G. 583 cm²
H. 1,510.81 cm²
I. 3,122.19 cm²
J. 1,771.56 cm²

Page 27

ON WICKENDS!

Page 28

A. 1,356.48 ft²
B. 460 cm²
C. 104 m²
D. 3,297 in²
E. 644 ft²
F. 5,200 cm²
G. 5,966 m²
H. 286 in²
I. 2,125 cm²
J. 480 cm²

Page 29

WALTON FAMILY

Page 30

SQUARE FEET